PARTS OF A PLANT

LEAVES

A Crabtree Roots Book

ALICIA RODRIGUEZ

Crabtree Publishing
crabtreebooks.com

School-to-Home Support for Caregivers and Teachers

This book helps children grow by letting them practice reading. Here are a few guiding questions to help the reader with building his or her comprehension skills. Possible answers appear here in red.

Before Reading:

- What do I think this book is about?
    - *This book is about leaves.*
    - *This book is about what leaves look like.*

- What do I want to learn about this topic?
    - *I want to learn how big leaves are.*
    - *I want to learn what colors leaves can be.*

During Reading:

- I wonder why...
    - *I wonder why leaves are green.*
    - *I wonder why plants have leaves.*

- What have I learned so far?
    - *I have learned that leaves can change color.*
    - *I have learned that leaves can be big or little.*

After Reading:

- What details did I learn about this topic?
    - *I have learned that leaves can be different colors, shapes, and sizes.*
    - *I have learned that most plants have leaves.*

- Read the book again and look for the vocabulary words.
    - *I see the word **leaves** on page 3 and the word **plants** on page 12.*

These are **leaves**.

Some leaves
are green.

Some leaves are brown, red, or yellow!

Some leaves are big.

Some leaves are little.

Most **plants** have leaves!

Word List

Sight Words

are
brown
green
little
most
red
some
yellow

Words to Know

leaves

plants

26 Words

These are **leaves**.

Some leaves are green.

Some leaves are brown, red, or yellow!

Some leaves are big.

Some leaves are little.

Most **plants** have leaves!

PARTS OF A PLANT

LEAVES

Written by: Alicia Rodriguez
Designed by: Rhea Wallace
Series Development: James Earley
Proofreader: Ellen Rodger
Educational Consultant: Marie Lemke M.Ed.

Photographs:
Shutterstock: Triff: cover; Nuk2013: p. 1; sek_suwat: p. 3, 14; Lumir Jurka Lumis: p. 5; Giovanni Love: p. 6-7; Jhanohiki: p. 9; cedesstuff: p. 11, 14; Tania Zbrodko: p. 13, 14

Crabtree Publishing

crabtreebooks.com 800-387-7650

 In Canada: We acknowledge the financial support of the Government of Canada through the Canada Book Fund for our publishing activities.

Printed in Canada/062024/CP20240604

Published in Canada
Crabtree Publishing
616 Welland Avenue
St. Catharines, Ontario
L2M 5V6

Published in the United States
Crabtree Publishing
347 Fifth Avenue
Suite 1402-145
New York, NY 10016

Hardcover	978-1-4271-4066-1
Paperback	978-1-4271-4072-2
Ebook (pdf)	978-1-4271-3333-5
Epub	978-1-4271-3393-9
Read-along	978-1-4271-4078-4
Audio book	978-1-4271-3912-2

Library and Archives Canada Cataloguing in Publication

Title: Leaves / Alicia Rodriguez.
Names: Rodriguez, Alicia (Children's author), author.
Description: Series statement: Parts of a plant | "A Crabtree roots book".
Identifiers: Canadiana (print) 2021017613X | Canadiana (ebook) 20210176148 | ISBN 9781427140661 (hardcover) | ISBN 9781427140722 (softcover) | ISBN 9781427133335 (HTML) | ISBN 9781427133939 (EPUB) | ISBN 9781427140784 (read-along ebook)
Subjects: LCSH: Leaves—Juvenile literature.
Classification: LCC QK649 .R63 2022 | DDC j575.5/7—dc23

Library of Congress Cataloging-in-Publication Data

Names: Rodríguez, Alicia, 1992- author.
Title: Leaves / Alicia Rodriguez.
Description: New York, NY : Crabtree Publishing Company, [2022] | Series: Parts of a plant - a Crabtree roots book | Includes index.
Identifiers: LCCN 2021014296 (print) | LCCN 2021014297 (ebook) | ISBN 9781427140661 (hardcover) | ISBN 9781427140722 (paperback) | ISBN 9781427133335 (ebook) | ISBN 9781427133939 (epub) | ISBN 9781427140784
Subjects: LCSH: Leaves--Juvenile literature. | Plants--Juvenile literature.
Classification: LCC QK649 .R634 2022 (print) | LCC QK649 (ebook) | DDC 581.4/8--dc23
LC record available at https://lccn.loc.gov/2021014296
LC ebook record available at https://lccn.loc.gov/2021014297